D1630174

A Journey Through Time

Written by **SELINA WOOD**

Illustrated by **RICHARD BONSON**

DK

A DORLING KINDERSLEY BOOK

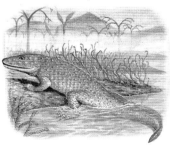
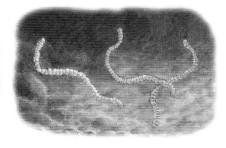

Dorling DK Kindersley

LONDON, NEW YORK, SYDNEY, DELHI, PARIS,
MUNICH, and JOHANNESBURG

Project Editor Selina Wood
Project Art Editor Lester Cheeseman
Art Editor Sheila Collins
Senior Editor Marie Greenwood
Managing Art Editor Jacquie Gulliver
Publishing Manager Jayne Parsons
Production Controller Jenny Jacoby
Picture Research Mollie Gillard, Jo Haddon
DTP Designer Almudena Díaz
Jacket Designer Dean Price

Archaeology Consultant Chris Thomas
Geology Consultant Margaret Carruthers

First published in Great Britain in 2001 by
Dorling Kindersley Limited
80 Strand
London WC2R 0RL

2 4 6 8 10 9 7 5 3 1

A CIP catalogue record for this book is
available from the British Library.

ISBN 0-7513-6162-3

Reproduced by Dot Gradations Ltd, Essex, UK
Printed and bound by L.E.G.O., Italy

See our complete catalogue at
www.dk.com

CONTENTS

A volcanic mud flow at Mayon, the Philippines, 1993, engulfs a community.

How the past gets buried

The Earth's surface is changing all the time. Decaying vegetation and eroded rock pile up in horizontal layers. In cities, the ground level is high because of the build up of demolished buildings and rubbish. Occasionally, natural disasters, such as earthquakes, disturb the layers, or volcanic eruptions speed up the layering process. A mud flow from a volcano can wipe out a town in minutes and set as hard as concrete. In other places, rock erodes away revealing a surface that can be millions of years old.

This stegosaur skeleton was reconstructed from fossil bones.

Fossils

The remains of living things that are preserved as rock are called fossils. In order for a life form to fossilize, it must be buried quickly, for instance by wind-blown sand or mud laid down by a river. Over millions of years, the bones are buried under more layers of mud which harden into rock. When this rock is eventually weathered away, the fossil is exposed.

Preservation or decay?

Very little of the past survives – most things are destroyed quickly or decay over time. Some substances decay more easily than others – the rate of decay depends on where the objects are buried. For instance, wood and leather usually only survive in waterlogged conditions, whereas pottery or bone can survive in many soils. The large amount of acid in peatbogs causes bones to disappear quickly, but preserves skin, hair, and insides.

This Roman sandal was preserved in waterlogged mud.

Dating

Many remains can be dated by following the principle that the deeper it lies, the older it will be. But sometimes a more precise method of dating living things (including wood and cloth) is by radiocarbon dating (above). The element radiocarbon (C-14) is present in all living things. When a living thing dies, the radiocarbon decays. Scientists know how quickly C-14 decays, and how much C-14 the animal contained when it was alive, so they can tell how long ago it died by measuring the level of radiocarbon in the remains.

The body was covered with layers of volcanic ash.

The ash gradually hardened into rock.

A cavity formed in the rock as clothes and flesh decomposed.

Fiorelli filled the cavity with liquid plaster.

Era	Period	Epoch	Millions of years ago (MYA)
Cenozoic	Quaternary	Holocene	0.01
		Pleistocene	1.6
	Tertiary	Pliocene	5
		Miocene	23
		Oligocene	36
		Eocene	58
		Paleocene	65
Mesozoic	Cretaceous		144
	Jurassic		208
	Triassic		245
Palaeozoic	Permian		286
	Carboniferous	Pennsylvanian (North America)	320
		Mississippian (North America)	360
	Devonian		408
	Silurian		438
	Ordovician		505
	Cambrian		570
	Precambrian		4,560

Geological time

Geologists (Earth scientists) have used rocks and fossils to create a timescale of the Earth's history. Geological time is divided into eras, periods, and epochs (above). If it were possible to find a place on Earth where layers of rock were constantly forming and never disturbed, they would lie in the same sequence as this timescale, with the oldest rocks at the bottom, and the newest at the Earth's surface.

Cast of the past

In the 18th-century excavations of Pompeii, Italian archaeologist Giuseppe Fiorelli discovered an ingenious method of reconstructing bodies that had been buried in the hardened volcanic ash. As the clothes and flesh of the victims decayed, a cavity (hole) was formed in the hardened ash. Fiorelli filled the cavity with liquid plaster, and when it had set, the ash was chipped away to reveal a perfect cast.

When the plaster set, the lava was chipped away to reveal a cast.

Today, resins are used to make cast as they are more durable than plaster.

600 BC Coins
Coins were first issued by the kings of Lydia, a state of western Anatolia (modern Turkey). By 560 BC the Greeks were also minting coins.

1000 BC Iron age begins
The smelting of iron ore began in Mesopotamia (modern day Iraq). They made implements and weapons out of this strong and hard-wearing metal.

Sculptures and vases show discus-throwing as a smooth, graceful movement. Sometimes the athletes were accompanied by musicians.

776 BC Olympic Games
Athletes from all over the Greek world came together to compete in games at Olympia, in honour of the god Zeus. Held every four years, the Olympic games went on to include wrestling, boxing, horse-racing, jumping, discus- and javelin-throwing, and chariot-racing.

The Chinese traded silk, ivory, jade, and spices for gold and silver from the Middle East. They travelled in groups or caravans for protection against bandits.

"Rainbow" bridges (so-called because of their shape) were built out of wood and bamboo.

The Chinese relied on water transport to carry goods all over China.

AD 1

85 BC Water-powered mill
The invention of mills powered by water from streams and rivers meant that people no longer had to use a pestle and mortar to grind grain for flour.

200 BC Silk roads
Trading routes between China and the West were busy with merchants exchanging goods (such as silk and horses), ideas, and technology.

447 BC Acropolis
The Greeks completed the Parthenon, a beautiful temple dedicated to the goddess Athena, built on the acropolis (hill fortress) above the city of Athens.

1800 BC Written laws
In Babylon, Babylonia (Mesopotamia), King Hammurabi introduced the first written laws. They were laws against family, civil, criminal, and commercial crimes and were inscribed on stone tablets.

1325 BC Tutankhamen
The body of Egyptian pharaoh Tutankhamen was mummified (preserved) before burial. The Egyptians, who believed in life after death, mummified bodies by extracting the main organs, sprinkling the bodies with a substance called natron and wrapping them in linen bandages.

Tutankhamen's burial mask was discovered in his tomb in 1922.

2350 BC Toilet
The first toilets were built in the Akkadian palace in Mesopotamia. They were simple pedestals with no water between them and the sewers – so they were probably very smelly!

2500 BC Tamed horse
Horses were first domesticated by humans for meat and skin. By 2500 BC the Mesopotamians were harnessing them to chariots and carts.

2500 BC Pyramids at Giza
The pyramids in Egypt were built to house the bodies of the pharaohs buried deep inside them. They were a great engineering achievement.

3000 BC

This ancient Mesopotamian tablet shows cuneiform text.

3000 BC Cuneiform writing
Scribes in Mesopotamia invented the first form of writing. They used a wedge-shaped reed to press symbols (representing objects) into tablets made of soft clay.

3000 BC Wheel
Simple wheels, made from three planks pegged together and cut into a circular shape, were constructed by the Mesopotamians. They were used to construct primitive carts.

AD 1 – 3000 BC

MAGNIFICENT CITIES

DIGGING BENEATH THE FOUNDATIONS of the road at Pompeii our journey takes us to earlier times and to far off places. We find traces of ancient civilizations – coins, statues, jewellery, tombs, weapons, even palaces. Time races back past milestones in human history – to China at the time of the first Emperor and his bustling city at Xianyang in 221 BC, and then to Ancient Greece, where politicians, philosophers, and priests participate in a religious festival on the Acropolis in Athens. We reach 2500 BC where, on the banks of the Nile, thousands of slaves drag colossal stone blocks across miles of sand to construct the Great Pyramid. And right at the dawn of civilization, scribes in Mesopotamia (modern Iraq) are discovering how to write.

The First Emperor's greatest project was the construction of the Great Wall, aimed at keeping out marauding tribes from the north. Many parts of the wall, including the section shown in this picture, were built at a later date.

CIVILIZATION IN CHINA

When Shih Huang-di became the First Emperor of all China in 221 BC he made his capital at Xianyang and set about building an organized society. He introduced systems to standardize coins, weights, and measures to encourage trade, and ordered that only one language be spoken. He also built canals and roads and organized a bureaucracy in which noble families were responsible for law and order in their own region.

Large bamboo sticks may have been used for construction because of their strength.

Peasants were forced to work on building projects, such as this canal.

Officials came to inspect the work. They would report progress back to the Emperor.

INTRODUCTION

THE GROUND BENEATH US holds the key to the history
of our planet. From bronze weapons found in ancient tombs,
to fossilized animals and plants discovered in sea cliffs, the
Earth's numerous treasures can tell us much about life
millions of years before we were born. As layer upon layer of
ground is formed from years of fallen vegetation, debris from
volcanic eruptions, discarded litter, and demolished buildings,
the secrets of the past become set in stone under our feet.

By digging deep into the Earth, we can begin to uncover its
secrets. But we can only dig so far. It is almost 6,400 km
(4,000 miles) to the centre of the Earth, but people have only
managed to drill a hole to a depth of about 15 km
(9 miles). They are unable to dig any further because of
kilometres of solid rock. Deeper still there is intense heat, so
hot that the rock becomes molten.

But let us imagine for a moment we can dig a deep tunnel
into the ground and begin excavating the history of the
Earth. The journey begins at the dawn of the 21st century. As
we travel back in time, we pass major events in history along
the way. And from a busy excavation site in Italy,
we begin our dig deep into the Earth.

1909 Motor car
American Henry Ford introduced the first mass-produced car, the Model T Ford. It was famous for its low cost and durability.

1903 Aeroplane
The *Flyer* took to the air. Built by the Wright brothers in the USA, it was the first powered aircraft to take off, fly under control of the pilot, and land safely.

1879 Electric light
American inventor Thomas Edison demonstrated the first incandescent lamp (electric light).

1839 Photography
Frenchman Louis Daguerre took some of the world's earliest photographs.

1876 Telephone
Scottish inventor Alexander Graham Bell made the first ever telephone call.

THE STREETS OF POMPEII

This is what the road would have looked like in AD 79, just before the volcano erupted. It is a busy shopping street with an ironmongers, a tavern with a bar, a laundry, several workshops and a bakery on the corner. Traders transport their goods in carts and shoppers make their way from stall to stall. The street is very dirty because people have dumped litter and rubbish, and because the drains empty straight out onto the road.

People lived over the shops in small apartments with no running water.

Large stepping stones were set into the roads so that people could cross without getting dirty.

The shops had open fronts with stalls where sellers placed their goods.

In the bakery, round loaves of bread were baked in a huge oven.

AD 2001

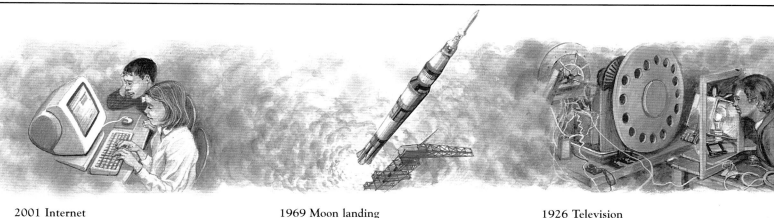

2001 Internet
People from opposite ends of the Earth can communicate from their personal computers (PCs) via the internet (a worldwide network).

1969 Moon landing
US spacecraft Apollo 11 landed on the Moon, and astronaut Neil Armstrong was the first person to step onto its rocky surface.

1926 Television
Scotsman John Logie Baird invented the television. It was one of several television systems under development at the time.

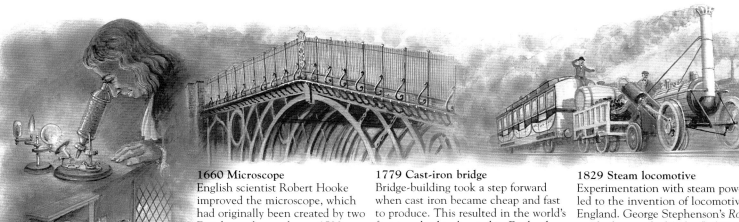

1660 Microscope
English scientist Robert Hooke improved the microscope, which had originally been created by two Dutch eyeglass-makers c.1590.

1779 Cast-iron bridge
Bridge-building took a step forward when cast iron became cheap and fast to produce. This resulted in the world's first iron bridge, located in England.

1829 Steam locomotive
Experimentation with steam power led to the invention of locomotives in England. George Stephenson's *Rocket* reached a speed of 50 km/h (30 mph).

1450 Printing press
German goldsmith Johann Gutenberg established the first printing press with movable type.

1347 Great plague
A great plague known as the Black Death, which was carried by rats, swept through Europe killing one in every three people.

1000 Gunpowder
Gunpowder was invented by Chinese scientists and revolutionized warfare.

AD 1

AD 1 Roman mining
Mining techniques originally used by the Ancient Greeks were adopted and improved upon by the Romans in the many mines throughout their Empire.

AD 100 Paper
The Chinese discovered that rags, tree-bark, or any fibrous material could be used to make paper. They kept their invention secret for hundreds of years.

AD 650 Windmill
The earliest windmills were built in Persia (modern-day Iran). The sails radiated from a vertical pole that turned with the wind, driving a pair of stones that ground the corn.

AD 2001 – AD 1

THE JOURNEY BEGINS

This mosaic, found inside the entrance hall of a villa in Pompeii, warned visitors to "beware of the dog". It tells us that pet dogs were used as guards in the same way as they are today.

THE AIR IS FILLED WITH THE CLATTER OF TROWELS scraping against rock, and the hum of people's chatter as they work. At an excavation site in Pompeii, southern Italy, archaeologists are busy uncovering what was once a busy Roman street lined with shops. In the year AD 79 catastrophe struck when Mount Vesuvius erupted, covering the town of Pompeii in volcanic ash and killing thousands of people. For hundreds of years, Pompeii lay buried and forgotten under layers of volcanic ash and vegetation, and farmers grazed their sheep and grew crops on the land. But since the 1700s, archaeologists have slowly dug away at the layers to reveal streets and houses, and everyday objects such as loaves of bread, mirrors, and pots and pans, miraculously preserved in ash. Archaeologists are still excavating today. Let us join them as they dig their way back in time to AD 79, and the busy streets of Pompeii.

EXCAVATION SITE

The archaeologists have dug down several metres and have uncovered most of this Pompeii street. Layer after layer of soil, rubble, or foundation is removed, and finds are put aside, labelled, and then taken to a laboratory for cleaning and dating. As they work, the archaeologists record everything they do by taking notes, drawings, and photographs. Knowing the exact location of every find will help them discover what life was like in the Roman town.

The site photographer takes photos at every stage.

A surveyor uses a level to take exact measurements of the excavated site.

The ancient ruins were once covered in layers of ash and soil, with pasture growing on the top.

An archaeologist brushes away ash from a large baker's oven that is perfectly preserved.

Finds are placed in a tray, ready to be taken to the laboratory.

The remains of a pot are found at the exact spot where it was dropped two thousand years ago.

Soil is delicately brushed away from the Roman stone.

The dimensions of the pavement are measured.

Every inch of the site is drawn in detail.

The Emperor was buried with an army of terracotta soldiers and horses. The tomb, which is presently being excavated, lies 40 km (25 miles) east of modern Xi'an.

Houses had timber frameworks with mud-brick outer walls and terracotta tile roofs.

The Emperor's palace was surrounded by a high wall with sentries who watched for intruders.

Scholars were encouraged to learn about medicine and agriculture. The Emperor forbade traditional subjects such as history and literature.

Fibre from silk worms was woven into cloth. The Chinese guarded the secret of how to make silk closely.

Civil servants kept records of who had paid their taxes. They wrote on silk or bamboo.

Merchants bought silk to sell for a high price in foreign lands such as Persia.

The standardization of coins, weights, and measures made buying and selling much easier.

Nobles, many of whom were famous soldiers, travelled in horse-drawn chariots. They wore silks and furs, had peasants to work their land, and sent their sons to school.

Music was played at the Emperor's court and in the city streets, particularly during festivals such as New Year. Pipes, gongs, and drums were common instruments.

3000 BC – 9000 BC

THE FIRST FARMERS

AS WE TUNNEL FURTHER DOWN, leaving the first Emperor's city far behind, the cities and towns of the ancient world are smaller. Temples and palaces are replaced by simple mud-brick buildings. Some settlements have walls built around them, and a few streets are lined with stalls where traders exchange pottery and other goods for food from local farmers. In small farming communities in Mesopotamia, workers sweat under a hot mid-day sun as they harvest wheat and barley and drive herds of cattle and goats into pens. We pass people learning how to weave flax fibres on a loom to make linen, and old tools made of wood and stone are cast away for stronger ones made of metal. Then we are back to 6000 BC and the first (man-pulled) plough is being tested in the fields. Tunnelling further back in time, we witness the very first farmers planting the first seeds that will lead to the spread of agriculture and the decline of the nomadic, hunting way of life.

Reed beds growing on the bank were cleared to make room for more fields.

Thousands of years ago, people discovered how to ferment fruits, vegetables, and grain to make alcohol. This wooden model, found in an Egyptian tomb, shows people making beer. The model represented workers that would accompany the dead person in the afterlife.

Water was drawn from the irrigation canals with a shadoof (a bucket fixed to a long rod on a pivot).

ON THE NILE

In 3000 BC, Egyptian farmers stored water from the river's summer floods in irrigation channels. This kept the soil damp and rich, so that their main crops – wheat, barley, and flax – grew well. In the autumn, the fields were ploughed, and the seed scattered and trodden in by herds of cattle. Then in spring, the crops were harvested and the grain stored in large bins. Fruit and vegetables were also cultivated, and animals, such as cattle, goats, and geese, were kept for meat, milk, and leather.

The Egyptians used wooden ploughs drawn by oxen. Seeds were scattered by hand behind the plough.

Seeds were trodden into the fields by animals.

Irrigation channels were built in the fertile agricultural plains.

11,000 YEARS AGO – 65 MYA (MILLION YEARS AGO)

A MAMMOTH TASK

THE AIR IN THE TUNNEL is getting colder and colder. Suddenly it opens out into a desolate landscape of mountains, glaciers, and evergreen trees. This is Earth during the last Ice Age (10,000–32,000 years ago). In an ice-bound valley, a mammoth wails in agony as it fights off marauding hunters, and a settlement of early people is busy making shelters out of mammoth bone and skin to protect themselves from the bitter wind. These are *Homo sapiens*, the first fully evolved humans, who use skill and ingenuity to counter the extreme cold.

As we venture further back in time we see the various stages of human evolution, as far back as *Proconsul*, a small, tree-climbing primate that is the ancestor of all modern apes and humans. All around us are bizarre-looking creatures, reminiscent of today's animals, but strangely different.

Mammoths were woolly, elephant-like animals of the Ice Age. This baby mammoth died at least 30,000 years ago. Its body was preserved in the frozen ground of a Siberian marsh, squashed flat by the weight of snow and earth above it.

THE ICE AGE

The Earth's climate alternates between cold periods (glacials) and warm periods (interglacials). During the last glacial, or Ice Age, living things had to cope with very harsh conditions. People were unable to live off the land because the vegetation had changed. Instead, they followed and hunted herds of mammoth, moose, or reindeer, setting up temporary camps near to their kill.

Weapons made from antlers were used to hunt wild animals.

Wood was collected from forests to make fire and to craft into weapons.

People hammered flints using heavy stones. These sharp-edged flints could then be used as spearheads.

Crafting a wooden pole to make a spear.

Shaping a spearhead into a sharp point.

Homo habilis skull

12,000 years ago First known pottery
Nomadic tribes in Japan started to use
clay dug from the earth to make pottery.
They made simple, round-based vessels.

13,000 years ago Dog domestication
Dogs were the first animals to be
domesticated. People in Israel tamed
wolves that scavenged around
human encampments.

17,000 years ago Cave paintings
Large animals, such as horses, bison,
and deer, were painted on the walls
of caves in Lascaux, southern France.

3.5 MYA *Australopithecine*
Our oldest ancestor evolved in
Africa. *Australopithecine* had a
small brain, walked on two legs,
and gathered fruits and
berries from trees.

2.5 MYA *Homo habilis*
"Handy man" was
short with curved arms.
They made stone tools,
and may have lived in
circular huts.

1.7 MYA *Homo erectus*
"Upright man" was an
efficient and organized
hunter who invented
new kinds of tools
and used fire.

**150,000 years ago
Neanderthals**
Homo sapiens neanderthalensis
lived in Europe and Asia.
They were probably the first
humans to bury their dead.

3.5–24 MYA Mammals dominate the Earth
The ancestor of modern apes and humans, *Proconsul*, used
hands to climb trees. A large carnivore, *Megistotherium*,
may have fed on elephant-like *Mastodonts*. A sabre-
toothed cat, *Megantereon*, hunted early antelope or deer.

24–34 MYA Largest-known mammal
One of the largest-known mammals, *Indricotherium*,
was 5.5 m (18 ft) tall and fed off leaves from the
tops of trees. Camel-like *Poebrotherium* roamed
the woodlands of North America.

55–65 MYA Early horse
Mammals began to multiply
and diversify. Most were small,
but by 55 MYA the dog-sized
Eohippus, an early ancestor
of the horse, had appeared.

34–55 MYA Mammals diversify
Fruits, grains, and grasses became abundant,
as did animal life. Species included
Icaronycteris (an early bat), *Diatryma*
(a flightless bird), and *Moeritherium*
(an early elephant).

34 MYA Marine mammals
Whales evolved from
hoofed land animals,
to strange, serpent-like
creatures, to toothed
and baleen whales.

32,000 years ago Bow and arrow
Hunters in the grasslands of Africa developed the bow and arrow to kill animals that were beyond the range of spears.

19,000 years ago Gathering cereal
In Israel, people gathered grain from wild cereals to crush it into flour to make simple bread.

37,000 years ago *Homo sapiens*
Fully modern humans (*Homo sapiens*) populated many parts of the world. They developed sophisticated speech and creative skills, and used ingenuity to survive the Ice Age.

Fine-stone hand axes were used by *Homo habilis* to chop through bone and meat.

The Egyptians grew large orchards that contained fig trees and vineyards.

Dates were gathered from the palm trees that grew abundantly along the Nile.

Goats were domesticated and kept for their milk.

Simple houses were built from mud-bricks that had been baked hard in the sun.

Farmers trod on grapes to extract juice in order to make wine.

Grain was stored in large bins made of mud-bricks.

Milk went sour in the heat, so most was made into cheese.

Grain was ground into flour, mixed with water to make dough, and then baked in an oven to make bread.

Reeds were woven into baskets and matting.

Farm tools, such as sickles, were made out of wood and flint blades.

The domestic cat originated in Ancient Egypt. It hunted the mice and rats that tried to feed from the grain bins.

65 MYA – 370 MYA

RULE OF THE REPTILES

WE FIND OURSELVES STANDING in a strange prehistoric lagoon. Small shrew-like animals run through lush ferns, crocodiles bask in the hot sun, and frogs splash in and out of the water. Suddenly a terrible roar echoes across the lagoon, branches on trees shake, and we glimpse a set of 60 huge dagger-like teeth and two flaring nostrils in the leaves. It is the head of a deadly *Tyrannosaurus* about to attack a bulky, plant-eating *Triceratops*, in a fight to the death. We have reached the Cretaceous period (65–144 MYA) and reptile-like animals, called dinosaurs, have been ruling the Earth for millions of years. Further back in time we watch one of the first dinosaurs, *Herrerasaurus*, running after its prey, giant insects flying in conifer forests, and primitive amphibians venturing out of water and adapting to life on land.

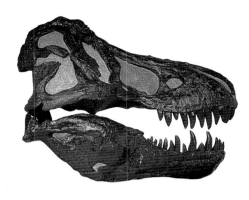

This fossil shows us that *Tyrannosaurus* had a huge skull with powerful jaws and fangs that stabbed deeply into flesh and bone, tearing out huge mouthfuls of meat. It could pick up victims and shake them violently apart.

Triceratops roamed in herds across the west of North America. Its horns and neck frill were used for attack and defence.

Tyrannosaurus was over 12 m (40 ft) long and might have weighed up to seven tonnes. It had strong legs to support its weight, but may not have been able to run fast.

The ostrich-like *Ornithomimus* was a swift runner. Its long tail helped it to balance and change direction quickly.

The tiny, clawed arms on *Tyrannosaurus* could bearly reach its mouth. It probably used them to hold its victims while it savaged them to death.

CRETACEOUS GIANTS

Dinosaurs ruled the planet for an astonishing 165 million years. From ancestors no bigger than dogs they evolved into a diverse group of carnivores and herbivores, the largest of which were several bus-lengths long. By the Cretaceous period, the flesh-eaters, such as *Tyrannosaurus*, had evolved into gigantic animals because they had to tackle enormous prey. It is likely that one of their victims was the sturdy, rhinoceros-like *Triceratops*, which fed on ferny vegetation.

65 MYA Meteorite
A large meteorite smashed into the Earth, causing vast clouds of dust to hide the Sun, cooling the climate. This may have led to the end of the dinosaurs, who had lived on the Earth for millions of years.

90 MYA *Icthyornis*
Icthyornis was the first bird that had strong wings and could fly for long distances. It was an agile creature, able to dive into water to catch fish.

100 MYA Flowering plants
Flowering plants similar to today's started to replace older kinds of vegetation, and pollinating insects appeared.

230 MYA The first dinosaurs
Dinosaurs evolved from primitive reptiles. One of the first was the small-to-medium-sized predator *Herrerasaurus*, which stood as high as an average man's waist.

245 MYA Mass extinction
Climate change may have been the cause of the largest mass extinction in the fossil record. Among the victims were the trilobites, who had thrived for 350 million years.

300 MYA Early reptiles
In the hot, dry climate, animals that could lay eggs on land rather than in water thrived. These early reptiles included long-legged *Hylonomus* and *Dimetrodon*, which may have used its "sail" to warm itself up.

320 MYA Coal swamps
Much of the land area on Earth was covered in swampy forest. Over millions of years the dead plant matter fossilized into today's coal.

370 MYA A fish out of water
A fish called *Eusthenopteron* had muscular fins that acted like legs. It may have been one of the first animals to move onto land in search of a greater variety of food.

350 MYA Thriving amphibians
Amphibians were the first vertebrates (animals with backbones) fully adapted to live on both land and water. *Ichthyostega* was able to breathe air, had legs for walking on land, and a large tail for swimming in water.

340 MYA Flying insects
Giant flying insects – including the dragonfly, which could grow to the size of a pigeon – flew in swampy, coniferous forests.

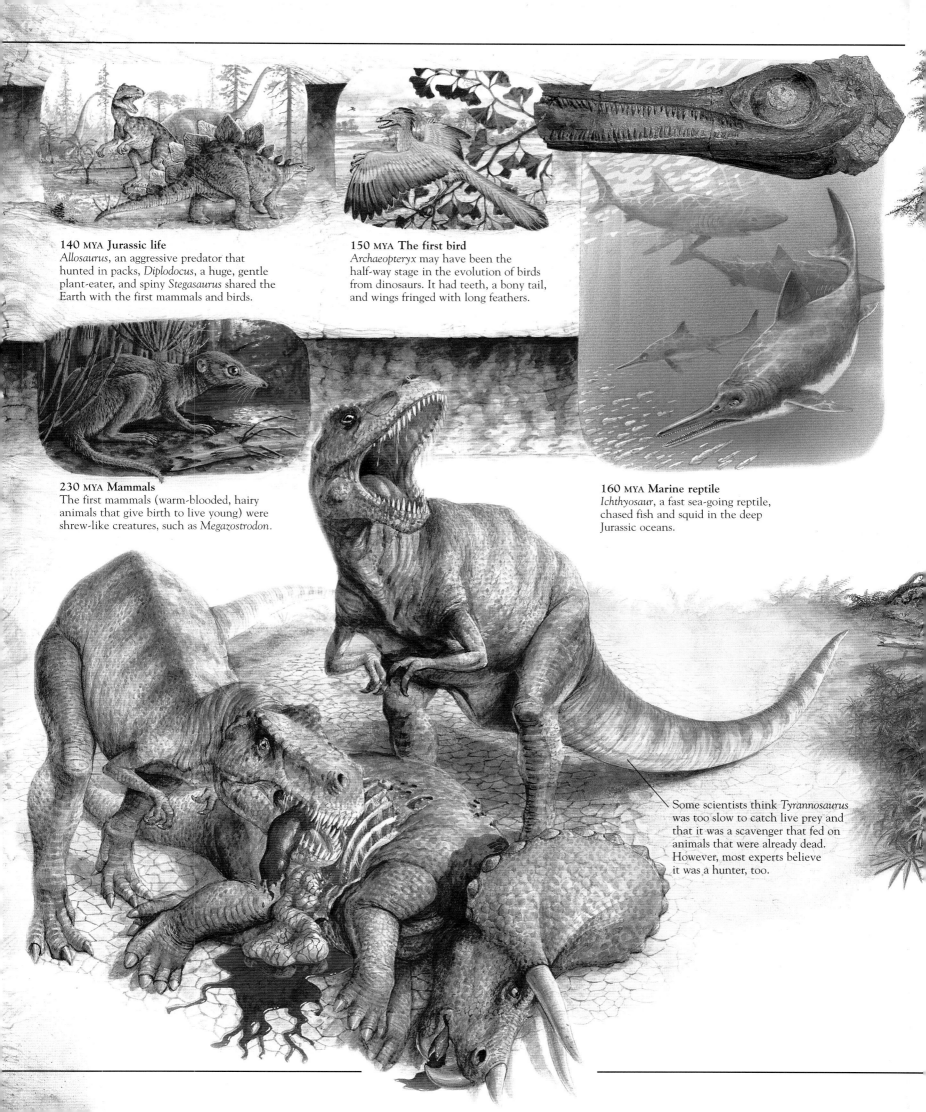

140 MYA Jurassic life
Allosaurus, an aggressive predator that hunted in packs, *Diplodocus*, a huge, gentle plant-eater, and spiny *Stegasaurus* shared the Earth with the first mammals and birds.

150 MYA The first bird
Archaeopteryx may have been the half-way stage in the evolution of birds from dinosaurs. It had teeth, a bony tail, and wings fringed with long feathers.

230 MYA Mammals
The first mammals (warm-blooded, hairy animals that give birth to live young) were shrew-like creatures, such as *Megazostrodon*.

160 MYA Marine reptile
Ichthyosaur, a fast sea-going reptile, chased fish and squid in the deep Jurassic oceans.

Some scientists think *Tyrannosaurus* was too slow to catch live prey and that it was a scavenger that fed on animals that were already dead. However, most experts believe it was a hunter, too.

Glaciers are ice sheets that form on mountains and move slowly forward under their own weight. They are common during Ice Ages.

Herds of reindeer roamed the icy landscape.

Cleaning a mammoth skin with scraping tools.

Mammoth bones were placed on top to weigh the skins down.

Specially treated skins were stretched over mammoth bones to build simple shelters.

Some skins were sewn together and made into clothes.

People roasted the mammoth's meat for food.

370 MYA – 3,800 MYA

LIFE IN THE SEA

THE DISTANT ROARS OF BATTLING DINOSAURS have long since faded away. We pass the first land animals, such as spiders and worms, and observe simple plant life taking hold on land. Then the tunnel merges into a blue underwater world of exotic sea life. Having now travelled back 400 million years, we have reached a time when life is evolving under water. Fierce hunters like *Dunkleosteus* chase schools of smaller shark-like fishes in and out of corals on the ocean floor. Some fishes have fins and jaws, others are flat with hard body shields, and a few have thick shells. The sea scorpions have fearsome pincers. Further back in time, soft, boneless sponges, jellyfish, and sea worms dominate the sea. Then, as we continue our journey, the water becomes emptier and emptier and the animals become so small that we cannot see them anymore.

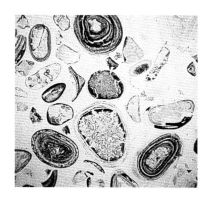

This thin section of chert (rock composed of fossils) shows the remains of some of the earliest forms of life – blue-green algae that grew in the sea from about 2,000 million years ago.

JAWS

The Devonian Period (360–408 mya) was the great age of the fish. Around 500 million years ago animals with backbones had appeared and 100 million years later many fish had fins and jaws. A backbone gave animals a support for softer body parts and anchorage for strong, body-bending muscles. Jaws were a great evolutionary advance because they enabled animals to bite off and cut up mouthfuls of food, rather than suck up small food particles from the water.

The ferocious jawed *Dunkleosteus* preyed on the primitive shark *Cladoselache*.

Round, flat-bodied *Gemuendina* fed on the seabed.

The primitive shark *Cladoselache* was an active predator.

370 MYA

The fern-like *Archaeopteris* may have been the world's first tree, forming the first forests. Fossils of this species have been found all over the world.

370–405 MYA The first plants and animals on land
Enough oxygen had built up in the atmosphere to support life, and plants took hold on land. As they became more widespread, animals such as worms, giant millipedes, and cockroaches evolved and fed upon them. In turn these creatures became the prey of meat-eaters such as scorpions and spiders.

The first vertebrates (animals with backbones) were fish, which were jawless.

480–600 MYA Starfish and sea urchins
Invertebrates (animals without backbones) thrived. Starfish and urchins fed on plankton and corals and sea lilies lived on the seabed, feeding on microscopic creatures.

Tentacled and sharp-eyed, the nautoloid was one of the top sea predators.

435–480 MYA Fan-shaped shells
During Silurian times, fish were jawless and had no fins, and animals with ribbed, fan-shaped shells, such as *Platystrophia*, were common.

Insect-like trilobites dominated the seas.

500–700 MYA Sea snails and brachiopods
Sponges, sea snails, and shelled animals called brachiopods slithered along the ocean floor. The bizarre *Hallucigenia*, with seven pairs of legs and seven pairs of tentacles, also inhabited the waters.

700 MYA Soft-bodied invertebrates
Simple, soft-bodied invertebrates such as jellyfish and circular-shaped worms, called *Dickinsonia*, first appeared. Leaf-shaped *Arborea* wafted on the seabed.

3,800 MYA

3,800 MYA The first living things
Microscopic bacteria were probably the first form of life, forming more than 3,500 million years ago.

3,200 MYA Stromatolites
In shallow oceans, simple plant life called algae grew in large dome-shaped mats. These mats have been preserved in stone as fossils, and are called stromatolites.

1,000 MYA Simple animals
The first animals appeared. Microscopic eukaryotes (multi-celled creatures) gradually evolved from the more primitive single-celled protozoa.

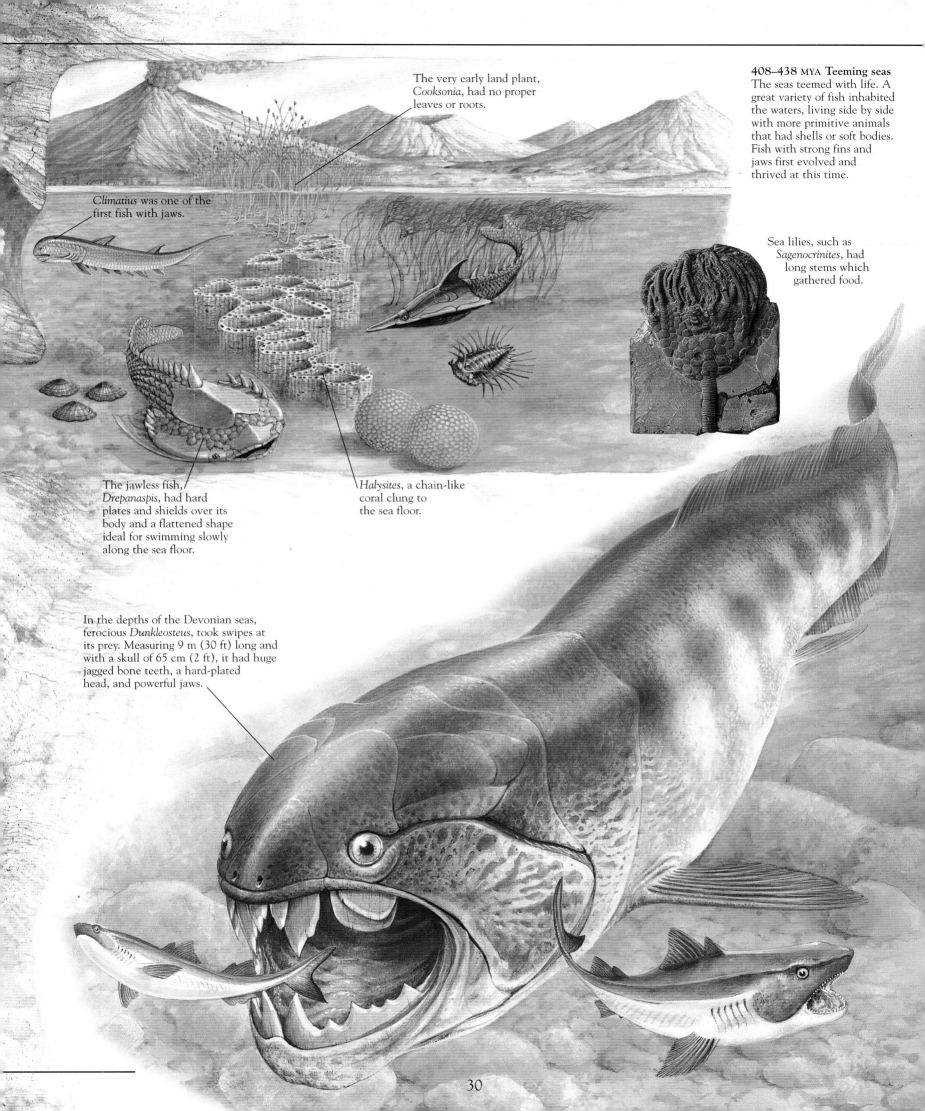

The very early land plant, *Cooksonia*, had no proper leaves or roots.

The seas teemed with life. A great variety of fish inhabited the waters, living side by side with more primitive animals that had shells or soft bodies. Fish with strong fins and jaws first evolved and thrived at this time.

Climatius was one of the first fish with jaws.

Sea lilies, such as *Sagenocrinites*, had long stems which gathered food.

The jawless fish, *Drepanaspis*, had hard plates and shields over its body and a flattened shape ideal for swimming slowly along the sea floor.

Halysites, a chain-like coral clung to the sea floor.

In the depths of the Devonian seas, ferocious *Dunkleosteus*, took swipes at its prey. Measuring 9 m (30 ft) long and with a skull of 65 cm (2 ft), it had huge jagged bone teeth, a hard-plated head, and powerful jaws.

Tall conifer trees covered many areas of Cretaceous Earth.

Parasaurolophus fed off a broad-leaved tree similar to today's oak.

Pterosaurs swooped over the Cretaceous landscape in search of fish. Some had a wing-span the width of a badminton court.

A *Euoplocephalus* used its heavy, clubbed tail to defend itself against a *Tyrannosaurus*.

The head-butting *Pachycephalosaurus* fought to achieve dominance in its social group.

Ferns and cycads grew abundantly by water.

The large *Hesperornis* bird had no wings – it was a swimming bird that chased and caught fish.

The duck-billed *Maiasaura* laid its eggs in mud nests and looked after its young. The newly hatched babies were as big as a human foot.

Crocodiles lay in wait for prey, ranging from fish to large vertebrates. They have changed little since prehistoric times.

3,800 MYA – 4,560 MYA

THE JOURNEY ENDS

OUR EXCAVATION HAS TAKEN US all the way back to the oldest rocks on Earth. We see the planet as it was 4,000 million years ago just as life was beginning. As we travel further back in time we see fragments of the Earth's crust forming in a sea of molten rock. By now we are immersed in billowing clouds of steam and gas. We pass landscapes with erupting volcanoes and huge craters formed by meteorites plummeting into the Earth. The air is filled with lightning flashes and explosions. All at once the tunnel comes to an abrupt end and we look down a steep shaft into a mass of hot, bubbling molten rock. We realize we are witnessing the birth of the planet 4.6 billion years ago, and our journey through time is complete.

Steam and other gases from erupting volcanoes formed the Earth's first atmosphere. Today there are about 1,300 active volcanoes on Earth, and innumerable extinct ones.

4,000 MYA
The Earth's early atmosphere was a smelly mixture of gases, too poisonous for most living things. Comets from space brought in some of these gases, including water, which condensed (turned to liquid) to form the oceans.

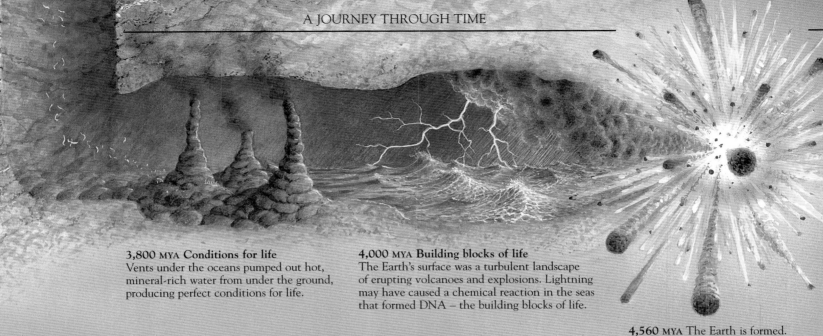

3,800 MYA Conditions for life
Vents under the oceans pumped out hot, mineral-rich water from under the ground, producing perfect conditions for life.

4,000 MYA Building blocks of life
The Earth's surface was a turbulent landscape of erupting volcanoes and explosions. Lightning may have caused a chemical reaction in the seas that formed DNA – the building blocks of life.

4,560 MYA The Earth is formed.

ROCKS AND MINERALS

When the Earth's molten surface cooled, it formed a solid surface, or crust, of rock. Over the years, the crust has become composed of many types of rocks, and almost all consist of one or more minerals (substances, such as quartz, formed naturally in the ground). There are three main rock types –igneous, sedimentary, and metamorphic – which form the building blocks of our landscape.

Metamorphic rocks, such as gneiss (left), are formed from igneous and sedimentary rocks that have been changed into new rocks by heat or pressure. These conditions can occur when mountains are formed.

Granite (below) and basalt are igneous rocks, formed when liquid rock from deep within the Earth cools and solidifies. The continents are made from granite.

Sedimentary rocks, such as sandstone and conglomerate (right), are formed from smaller pieces of other rocks that have eroded away over many years.

MOVING PLATES

The Earth's crust is broken up into about 15 vast, thick pieces of rock known as plates. These slabs fit together like a huge jigsaw. Like pieces of ice on a partially frozen river, they are constantly moving against each other. At their boundaries, the plates may be colliding, pulling apart, or sliding past each other. These different types of motion build mountains, cause earthquakes and volcanoes, and create deep sea trenches. The plates' movement is thought to be caused by the movement of hot rock deep in the Earth below.

When plates collide, one sometimes rides over the other, forcing it down into the mantle. This often occurs where thick continental plates ride over thinner oceanic plates, forming deep trenches in the ocean. Volcanic eruptions occur frequently.

Volcanoes occur where intense pressure forces gases and magma to rise from the mantle through an opening in the crust.

Rocks begin to melt when a plate is forced down into the mantle.

Earthquakes often occur where two plates slide and judder past each other.

Where two continental plates collide, the rocks pile on top of each other to form mountain ranges.

Mid-ocean ridges and steep valleys on land occur where two plates pull apart, and molten rock from the mantle rises to form new crust.

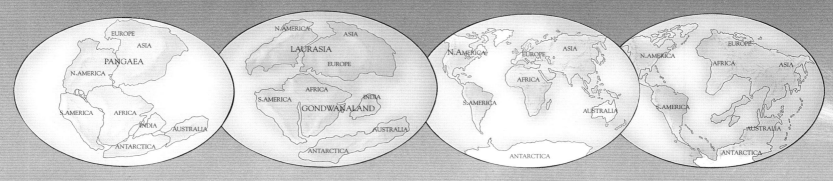

250 MYA
The continents were clustered together in a giant super-continent called Pangaea, from the Greek meaning "all lands".

135 MYA
Pangaea split into two main land masses, Laurasia and Gondwanaland. North America and Europe split apart.

Today
The Americas have moved away from Europe and Africa. India has joined Asia, and Australia and Antarctica have split apart.

150 MY in the future
Africa may split in two, and the larger section may drift north to join Europe. Antarctica may rejoin Australia.

INSIDE THE EARTH

The Earth today consists of four main layers – an inner core, outer core, mantle, and crust. The heat from the core causes material in the molten outer core and mantle to circulate. These cycles are known as convection currents. Convection in the outer core creates a magnetic field (force) that helps to shield the Earth from radiation from the Sun. Convection in the mantle moves the crust's plates around.

The crust is made of mainly basalt and gabbro rock and is 7–40 km (4–25 miles) thick (the thinnest part being under the sea).

The mantle is about 2,800 km (1,700 miles) thick. Although it is very, very hot, the pressure of the rocks above keeps it solid.

The outer core of molten iron and nickel is about 2,300 km (1,400 miles) thick.

The solid inner core of iron and nickel is about 2,400 km (1,500 miles) in diameter, and is extremely hot – 4,000° C (7,000°F). The heat is thought to be left over from when the Earth formed.

Convection current

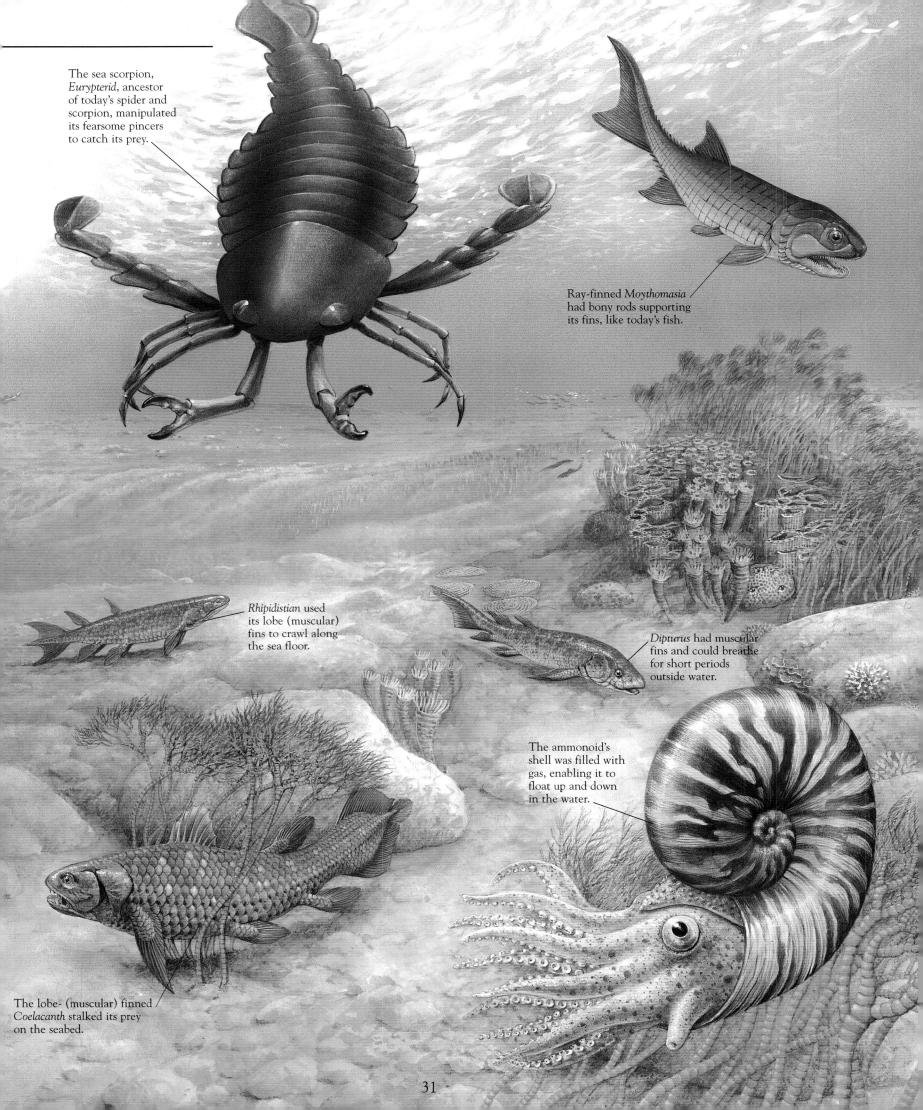

The sea scorpion, *Eurypterid*, ancestor of today's spider and scorpion, manipulated its fearsome pincers to catch its prey.

Ray-finned *Moythomasia* had bony rods supporting its fins, like today's fish.

Rhipidistian used its lobe (muscular) fins to crawl along the sea floor.

Dipturus had muscular fins and could breathe for short periods outside water.

The ammonoid's shell was filled with gas, enabling it to float up and down in the water.

The lobe- (muscular) finned *Coelacanth* stalked its prey on the seabed.

THE MOON

The Moon is slightly younger than the Earth and revolves around the Earth as it orbits the Sun. Scientists think that it was formed from vast amounts of matter that were thrown into space when a small planet collided with the young Earth.

4,560 MYA
About 4,560 million years ago, a dense cloud of gas and dust in space contracted to form the Sun. Other matter in the cloud formed solid lumps of ice and rock, and these joined together to form a rocky planet.

The Earth's first rocky skin was ripped apart by the impact of huge meteorites (loose rock fragments from space). The build-up of heat from the explosions caused the newborn Earth to melt. Iron and nickel in the rocks sank to form the Earth's core, while oceans of liquid rock floated on the surface.

This meteorite fell in Leicestershire, England, in 1965. The meteorite is 4,600 million years old and was formed at the same time as the planet, but in a different part of the solar system.

Earth had a thin, fragmentary crust that continually cracked and remelted – solid crust sank back into the hot interior to melt again, while new crust grew and cooled. As the Earth cooled off, pressure from gases and liquid rock in the interior created volcanoes on the surface.

EARTH'S PLACE IN THE SOLAR SYSTEM

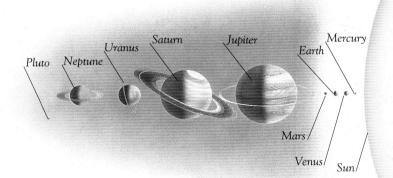

Pluto · Neptune · Uranus · Saturn · Jupiter · Earth · Mercury · Mars · Venus · Sun

The Earth is one of nine planets that orbit the Sun. The planets were formed from smaller rocks called planetesimals, which emerged from the gas and dust left over after the Sun was formed. Earth is the fifth largest planet and is the only one known to support life. This may be due to the water in the oceans, oxygen in the atmosphere, and its ideal location from the Sun, making it neither too hot nor too cold.

FAMOUS DISCOVERIES

NOW THAT WE HAVE COMPLETED our excavation of the Earth's history, we can put down our spades, take a rest, and dwell on some of the most fabulous treasures that have been dug up from under the ground. These amazing finds have changed the way we think about the Earth's history. There are *Iguanodon* bone fossils, among the first signs of evidence that dinosaurs once ruled the planet. There are also remarkable treasures from Tutankhamen's tomb, revealing the wealth and religious rites of the Ancient Egyptians. And finally, we can marvel at the frozen man who saw the light of day after a staggering 5,000 years.

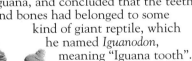

IGUANODON
TEETH

IGUANODON
BACKBONE

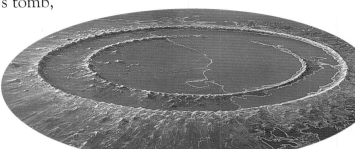

GIANT BONES

Although dinosaur remains have existed for millions of years, people did not know what they were until the 19th century. In 1820, an English doctor, Gideon Mantell, stumbled across some large teeth and bone fossils embedded in stone. After much research, Dr. Mantell realized that the bones were similar to that of an iguana, and concluded that the teeth and bones had belonged to some kind of giant reptile, which he named *Iguanodon*, meaning "Iguana tooth".

ANCIENT CITY OF TROY

In the 8th century BC, the Greek poet Homer told of an ancient city called Troy that was at war with Greece. Following Homer's clues as to where the city was, German archaeologist Heinrich Schliemann set out to find Troy in the 1870s. He unexpectedly succeeded, exposing the remains of old houses in an ancient city dating from c.1200 BC.

TERRACOTTA ARMY

The first Chinese emperor, Shih Huang-di, who reigned in the 3rd century BC, designed his own tomb, one of the greatest ever built. It was a vast, underground palace with jewelled roofs and rivers of flowing mercury. Thousands of life-size terracotta warriors with weapons guarded the entrance of the tomb. The tomb was dramatically discovered by Chinese peasants digging a well in 1974.

CHICXULUB CRATER

In the early 1990s, geologists discovered a 180-km (112-mile) wide crater buried deep under rock on the Mexican coast. They think it was formed when a comet or asteroid smashed into the Earth 65 million years ago – exactly when the dinosaurs disappeared – and that it may have been the cause of the dinosaurs' mass extinction. This artist's impression of the crater shows how it might have looked soon after the impact, with a massive circle of mountains.

SUTTON HOO
HELMET

SUTTON HOO

Excavations of a 7th-century Anglo Saxon grave at Sutton Hoo in England in the late 1930s revealed many artefacts including a stunning helmet made of bronze and iron. Before this discovery, people thought that Anglo Saxon metal work was much more primitive.

The Ancient Egyptians believed the coffin was a house for the dead person's spirit.

TUTANKHAMEN'S TOMB

Tutankhamen was an Ancient Egyptian pharaoh (king) who died at the young age of 18 in 1327 BC. Little is known about his life, but his tomb is extremely famous. It was rediscovered by English Egyptologist Howard Carter in 1922. The rooms were crammed with furniture, clothes, a chariot, and weapons, and the pharaoh's mummy lay within a nest of three golden coffins, one inside the other. Because the tomb was relatively untouched by tomb robbers, it reveals much about the Ancient Egyptians. It shows us that the pharaohs had immense power and wealth, and that they were buried with everything they would need in the after-life.

The second of Tutankhamen's coffins is made of gilded wood.

The coffin is inlaid with red and turquoise glass and blue pottery.

MACHU PICCHU

An abandoned city high up in the Andes in Peru was covered in dense vegetation and remained unknown until its excavation by the American archaeologist Hiram Bingham in 1912. The 300-year-old city belonged to the Incas, a South American people who ruled an empire that extended from modern Ecuador to Chile. Experts believe Machu Picchu was the site of a magnificent palace and several temples.

CLEOPATRA'S PALACE

Since 1992, French archaeologist Franck Goddio and a team of divers have been excavating the harbour at Alexandria, Egypt, hoping to find the royal quarter where Cleopatra VII of Egypt (51–30 BC) lived. The site of the royal quarter was flooded after an earthquake in the 14th century. So far the underwater excavators have found a variety of statues and artefacts.

ICE MAN

In 1991, two hikers in the Italian alps discovered the frozen remains of a man under the snow. On close examination, experts reported that the body was over 5,300 years old. The low temperatures had kept the body so well preserved that the researchers could tell the age of the man (46) and what his last meal had been – meat and a type of hard bread.

A tattoo is still visible on the ice man's arm.

INDEX

ACKNOWLEDGMENTS

Dorling Kindersley would like to thank:
Chris Bernstein for the index;
Carole Oliver, Robin Hunter, and
Polly Appleton for design assistance;
Lee Simmons for editorial assistance.

Picture credits
The publisher would like to thank the following for their
kind permission to reproduce their photographs:
c=centre; b=bottom; l=left; r=right; t=top; a=above

Ashmolean Museum:17bc.; **Museo Archaeologicao di
Nationale:** 8tl. **Museo Archaeologicao di Pompei:** 11br.
British Museum, London: 13bc, 14tc. **Corbis:** 12tl. **Robert
Harding Picture Library:** 13cr, 36bl, 37cl; Chris Rennie 37tr;
Ken Wilson 36c. **Museum of London:** 11cla, 17tl. **Popperfoto:
Reuters:** 37cr. **Powerstock Photolibrary/Zefa:** Ian Lishman
35tl. **Rex Features:** 11tl. **Science Photo Library:** D Van
Ravenswaay 36cra; David Weintraub 32tl; James King-Holmes
11ca. **Sygma:** George de Keerle 37br.
All other images © Dorling Kindersley.
For further information see: www.dkimagea.com.